LAYERS OF THE OCEAN

KlevaKids

Published by KlevaKids.com Inc

New York, New York

For more titles,visit our website at www.klevakids.com

KlevaKids

KlevaKids

Presents

LAYERS OF THE OCEAN

The ocean layers are presented in the context of zones, including the sunlight zones context.

The above statements are purely informational. The purpose of the book is not to teach, but to expose concepts. Read the book to a child and just allow the story to be enjoyed.

It was a gas cloud

beyond any doubt.

Who could have known

what was to come about?

When the gases condensed,

it made more than dust.

Earth formed with a hard shell

called the crust.

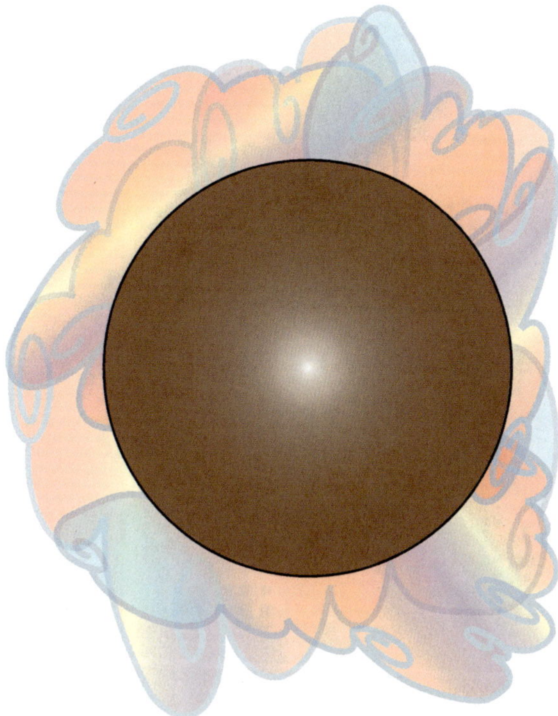

Hills and holes appeared.

They looked like dents.

To release the pressure inside,

the earth formed vents.

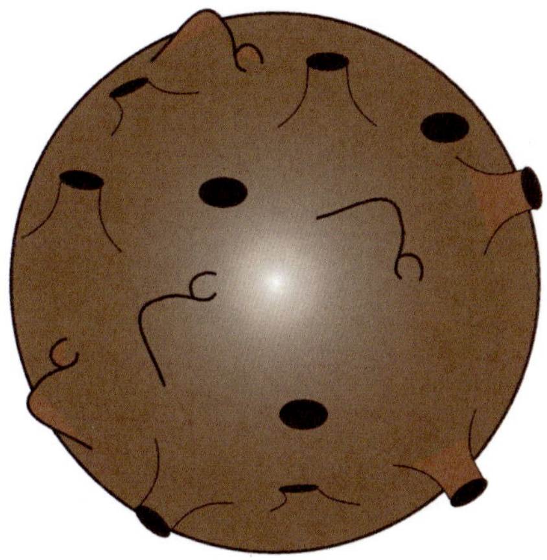

Released from the vents

was lots of steam

that made a cloud so large

it was like a dream.

The cloud made rain

to create an ocean.

It formed the hadalpelagic zone

in one swift motion.

Hadalpelagic Zone

Out of the cloud

came more rain.

It made the abyssopelagic zone

without a strain.

Abyssopelagic Zone
Hadalpelagic Zone

Just how much rain

would be enough?

It made the bathypelagic zone

with just a little puff.

Bathypelagic Zone
Abyssopelagic Zone
Hadalpelagic Zone

The rain continued.

There was more.

Making the mesopelagic zone

was just a little chore.

Mesopelagic Zone
Bathypelagic Zone
Abyssopelagic Zone
Hadalpelagic Zone

Just when it seemed

that the rain would never stop.

the epipelagic zone

was the last one on top.

Epipelagic Zone
Mesopelagic Zone
Bathypelagic Zone
Abyssopelagic Zone
Hadalpelagic Zone

When the rain stopped,

the earth was quite grand.

It was covered by a lot of water

and not much land.

The sun was watching

from quite a distance.

It felt the names were too complex

and provided assistance.

The sun thought the layers

should be three and no more.

The aphotic zone without any light

is near the ocean floor.

Aphotic Zone

The middle was the disphotic zone

where some things like to swim.

There is not much light there.

In fact it's very dim.

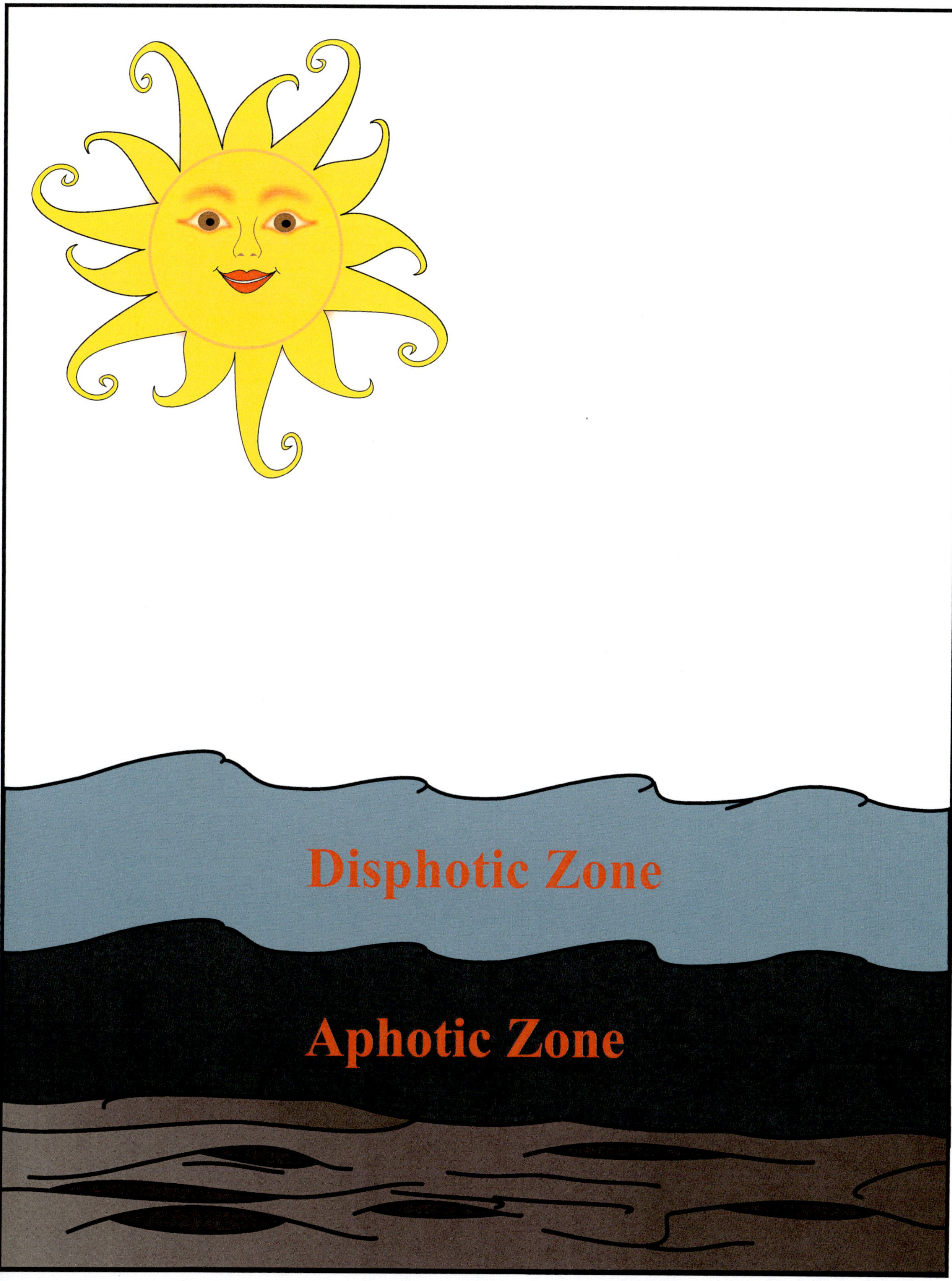
Disphotic Zone
Aphotic Zone

The euphotic zone is at the top

and the sun likes to boast

that's where the light

is in the water the most.

Euphotic Zone
Disphotic Zone
Aphotic Zone

The ocean said to the sun and cloud that these names are too confusing. The sunlit, twilight, and midnight zones are the names we will be using.

THE END

Sunlit Zone
Twilight Zone
Midnight Zone

PRONUNCIATION GUIDE

Epipelagic (e-pi-pe-la-jik)

Mesopelagic (me-za-pe-la-jik)

Bathypelagic (ba-thi-pe-la-jik)

Abyssopelagic (a-bis-o-pe-la-jik)

Hadalpelagic (hay-del--pe-la-jik)

Euphotic (yoo-foh-tik)

Disphotic (dis-foh-tik)

Aphotic (a-foh-tik)

Made in the USA
Las Vegas, NV
24 June 2026

49682416R00021